KITCHEN
MATH

SUSAN BRENDEL

Cover illustration by Eileen Stamper

J. WESTON
WALCH
PUBLISHER
PORTLAND, MAINE

1   2   3   4   5   6   7   8   9   10

ISBN 0-8251-2881-1

Copyright © 1988, 1997
J. Weston Walch, Publisher
P.O. Box 658 • Portland, Maine 04104-0658

Printed in the United States of America

CONTENTS

Planning Ahead

Shopping for Food

Stretching Food Dollars

Using Measurements

Preparation Time

Kitchen Supplies

To the Teacher

Objectives

1. To provide students practice in using the basic math operations (addition, subtraction, multiplication, and division) and applying them in real-life situations that involve the selection, purchase, and preparation of foods and the purchase of other kitchen necessities.

2. To help students develop problem-solving skills with an emphasis on knowing when to apply which math operations. (To this end, you may wish to employ the exercises for optional practice using the hand-held calculator.)

3. To provide useful and flexible classroom materials to help teachers develop and evaluate specific math competencies.

4. To make students more aware of some of the practical aspects of shopping, meal planning, and working in the kitchen.

Content

The units in this set include: Planning Ahead, Shopping for Food, Stretching Food Dollars, Using Measurements, Preparation Time, and Kitchen Supplies. A complete list of individual topics appears in the Contents.

Grade Levels

Kitchen Math has been developed for students in middle school through high school, especially those who are having difficulty with math. The set is designed to be useful in any classes at those levels where fundamental math, consumer, and homemaking skills are covered.

The math skills are approximately fourth- to sixth-grade level. The reading level is grade four. The interest level is age twelve to adult.

Motivational Aspects

1. Most secondary school students and adults are, to varying degrees, involved with selecting and buying food and working in the kitchen. Students will feel that the content of *Kitchen Math is* relevant to their personal needs.

2. The worksheets are interesting and attractive, an appealing alternative to the standard textbook pages students have seen before.

3. The review pages at the end of each unit include a motivational self-checking device that students will enjoy.

4. The skills practiced in the series are basic enough so that many students will experience success in math, perhaps for the first time.

How to Use This Set

The materials are designed to require little teacher direction. Most of the masters are largely self-explanatory. Some students may require individual explanation of the directions and specific review of the topic or process presented.

For your convenience in the classroom, the set incorporates the following features:

1. An answer key permits easy checking, providing answers for all worksheets, review pages, and test sheets.

2. A vocabulary list points out specialized terms related to the subject matter for specific masters.

3. The pretests and posttests provide a ready-made evaluation tool for you.

The set may be used with individuals, small groups, or an entire class. In some instances, students could progress through the material at their own rates, thus providing an individualized instruction program. In other situations, *Kitchen Math* would make an excellent group-instruction tool.

Since the masters are self-contained, you could use pages from *Kitchen Math* as practice sheets for specific skills. Rather than having the whole class complete every master, you could give each student only the masters pertaining to those skills in which the student is deficient. Skills practiced on each master are easily identified using the Math Skills chart on page *viii* of this book.

Evaluating Student Progress

Kitchen Math contains pretests and posttests for both computational and application problems similar to those which students will encounter in the exercises.

Pretests

• The Computation Pretest (Masters #1 and #2) tests the basic computation skills included in *Kitchen Math*. It may be used to assess a student's readiness to do the practice problems in *Kitchen Math*. Or it may be used to identify specific areas in which a student's skills are weak.

• The Application Pretest (Masters #3–5) tests a student's ability to use basic computational skills in practical situations. One problem representative of those presented on each master in *Kitchen Math is* included in the pretest. Results from this pretest will indicate to you which masters a student may be able to do independently and which may require more teacher direction.

Posttests

- The Computation Posttest (Masters #44 and #45) allows you to evaluate a student's progress in computational skills when compared to the score on the pretest (Masters # 1 and #2). Areas needing further work will also be clear from the test results.

- The Application Posttest (Masters #46–48) allows you to evaluate a student's progress in applying basic math skills to solve practical problems in *Kitchen Math* as compared to the score on the pretest (Masters #3–5).

Each numbered problem on the Application Posttest correlates with the problem of the same number on the Application Pretest. Thus, you may directly compare results and make inferences as to any areas you may need to reteach.

The chart below indicates the master from which each problem on the Application Pretest and/or Posttest was taken:

Problem	Master	Problem	Master	Problem	Master
1	6	12	19	23	32
2	7	13	20	24	33
3	8	14	21	25	34
4	9	15	22	26	35
5	10	16	23	27	36
6	12	17	25	28	38
7	13	18	26	29	39
8	14	19	27	30	40
9	15	20	28	31	41
10	16	21	29	32	42
11	17	22	30		

Math Skills Presented

Emphasis is on the application of basic math skills; that is, knowing when to apply which math operations. Most masters require addition, subtraction, multiplication, or division of whole numbers or decimals (dollars and cents). Some of the masters involve working with fractions and mixed numbers, ratios, percents, and units of measure, including time.

The following chart gives a specific breakdown of math skills practiced on each master. The review sheets cover multiple skills and so are not listed here.

Worksheets	Whole Numbers				Money / Decimals				Fract.	Time	%
	+	−	×	÷	+	−	×	÷			
PLANNING AHEAD											
6. Counting Calories	•										
7. Nutrition Facts		•	•								•
8. Figuring the Cost of One Meal					•	•					
9. The Food Budget					•	•					
10. Planning a Cookout			•	•							
SHOPPING FOR FOOD											
12. The Shopping List					•		•				
13. Count Your Change						•					
14. Fresh Produce by the Pound							•				
15. Read the Label							•				
16. How Much for One?								•			
17. Buying Baby Food				•			•				
STRETCHING FOOD DOLLARS											
19. Saving Money with Coupons						•					
20. Comparing Store Prices					•	•					
21. Comparing Brands						•	•				
22. Unit Pricing								•			
23. Buying Large Quantities				•		•	•				
USING MEASUREMENTS											
25. Equal Measures			•								
26. Combining Liquid Ingredients	•										
27. Combining Dry Ingredients									•		
28. Figuring Leftovers									•		
29. Adjusting a Recipe			•	•					•		
30. Ratios in the Kitchen									•		
PREPARATION TIME											
32. Cooking Times										•	
33. Microwave Cooking Times										•	
34. Serving a Meal on Time										•	
35. Cooking in Batches			•	•						•	
36. How Long to Cook Meat?			•	•						•	
KITCHEN SUPPLIES											
38. Buying Cleanup Supplies on Sale					•	•					
39. Buying Kitchen and Table Linens					•	•	•	•			
40. Cost of Tools for Cooking and Serving					•		•				•
41. Appliance Discounts						•	•				•
42. Buying Appliances on Credit					•			•			

Vocabulary List

The reading level of *Kitchen Math* is fourth grade. However, some specialized vocabulary and abbreviations are required because of the subject matter. These terms are listed below. You may wish to explain these terms before having your students do the work for those masters.

Master #6 calories, energy, the various foods listed

Master #7 nutrition, serving, calories, iron, dietary fiber, carbohydrates, the other nutrients shown on the "Nutrition Facts" label

Master #8 the various foods listed

Master #9 budget, groceries

Master #10 cookout, menu, oz (ounce), the various foods listed

Master #12 shoppers, avoid, oz (ounce), lb (pound), gal (gallon), doz (dozen), pkg. (package), the various foods listed

Master #13 change, groceries, cash register, computer, cashier, tax, taxable, nontaxable, froz. (frozen), cash tdr. (cash tendered, or paid), the various foods listed

Master #14 produce, in season, produce section, priced, lb (pound), per lb (for one pound), the various foods listed

Master #15 priced, net wt. (net weight), lb (pound), per lb (for one pound)

Master #16 yogurt, quantity, round off, the various foods listed

Master #17 budget, servings, oz (ounce), formula

Master #19 coupon, cashier, worth, expiration date

Master #20 compare, convenience (variety) store, grocery store, supermarket, shoppers, pkg. (package), the various foods listed

Master #21 compare, brands, name brand, store brand, generic (or "no-name") brand, the various foods listed

MASTERS

Name _____ Date _____

Computation Pretest: Part One

Directions: See how many problems you can do. You may need another sheet of paper to work the problems.

Add

1.　　435
　　+　15
　　————

2.　　375
　　　60
　　+ 485
　　————

3.　　$2.30
　　$ .45
　　+ $1.20
　　————

4.　　$ 1.79
　　$　.16
　　+ $19.90
　　————

5.　　$49.95
　　+ $ 2.09
　　————

6.　$\frac{1}{2}$ + $\frac{1}{2}$ = _____

7.　　$3\frac{1}{4}$
　+　　$\frac{1}{2}$
　————

Subtract

8.　　3600
　　− 2850
　　————

9.　　$8.00
　　− $3.00
　　————

10.　　$1.89
　　−　.15
　　————

11.　　$15.00
　　− $12.06
　　————

12.　　$35.00
　　− $33.97
　　————

13.　　$.58
　　− $.49
　　————

14.　$2\frac{1}{2}$
　−　$\frac{1}{2}$
　————

15.　$\frac{1}{2}$
　−　$\frac{1}{4}$
　————

Kitchen Math

Computation Pretest: Part Two

Directions: See how many problems you can do. You may need another sheet of paper to work the problems.

Multiply

16. $\begin{array}{r} 24 \\ \times\ 6 \\ \hline \end{array}$

17. $\begin{array}{r} \$1.50 \\ \times\ \ \ 7 \\ \hline \end{array}$

18. $\begin{array}{r} \$\ .79 \\ \times\ \ \ 4 \\ \hline \end{array}$

19. $\begin{array}{r} \$2.44 \\ \times\ .25 \\ \hline \end{array}$

20. $\begin{array}{r} \$\,35 \\ \times\ .20 \\ \hline \end{array}$

21. $\$299 \times 10\% = $ _____

22. $\frac{1}{2} \times 6$ _____

Divide

23. $6\,)\overline{48}$

24. $8\,)\overline{\$3.20}$

25. $60\,)\overline{120}$

26. $\frac{1}{2} \div 2 = $ _____

Kitchen Math

Name _____ Date _____

Application Pretest: Part One

(For Masters #6–17)

Directions: See how many problems you can do. You may need another sheet of paper to work the problems.

---

1. A lamb chop has 140 calories.
   A potato has 90 calories.
   A glass of milk has 165.
   What are the total calories?

2. One-quarter cup of raisins gives you 130 calories. Suppose you should get 2300 calories in one day. How many calories will come from other foods?

---

3. Gina bought these foods:
   Roast beef $ 10.50
   Tomatoes $ .40
   Bread $ 1.00
   What was the total cost?

4. Rob's food budget is $25. He has spent $17. How much does Rob have left?

---

5. There are 16 students. Each wants 4 pickles. How many pickles do they want in all?

6. A bag of 8 hamburger buns costs $1.39. Ella wants 3 bags. What will she pay?

---

7. Don's groceries cost $9.50. He pays with a $10 bill. How much change will he get?

8. Onions are $.39 per pound. How much will Mike pay for 6 pounds?

---

9. A package of meat weighs 3.5 pounds. The price is $1.50 per pound. What is the price of this package?

10. Jack sees 3/$.99 marked on canned corn. How much will he pay for just one can?

---

11. A baby bottle holds 8 ounces. A jar of juice contains 32 ounces. How many baby bottles will one jar fill?

Application Pretest: Part Two

(For Masters #19–30)

Directions: See how many problems you can do. You may need another sheet of paper to work the problems.

12. Juice costs $1.19. Lee has a coupon that says 20¢ off. What will he pay for juice?	13. Cereal is $1.83 at one store. It is $2.09 at another store. How much can you save at the less expensive store?
14. One brand of bacon is $1.89 per pound. The other brand is $1.79. How much can you save buying 3 pounds of the lower-priced brand?	15. Which is the better buy? a. 64 oz for $2.56 b. 32 oz for $1.60
16. A 10-lb bag of flour costs $3.69. A 5-lb bag costs $1.55. How much can you save if you buy the big bag instead of 2 small bags?	17. There are 8 ounces in one cup. How many ounces are in 5 cups?
18. You combine 650 mL of milk and 150 mL of cream. How much liquid do you have in all?	19. You mix 1¼ cups of rice and 1¼ cups of beans. How much mixture do you have in all?
20. Mitch has 2 pounds of rice. If he uses ½ pound, how many pounds are left?	21. A recipe uses 4 potatoes. How many potatoes make a half recipe?
22. One orange makes 4 ounces of juice. How many oranges make 20 ounces?	

Application Pretest: Part Three

(For Masters #32–42)

Directions: See how many problems you can do. You may need another sheet of paper to work the problems.

23. Amy starts to bake a cake at 2:15. It bakes for 30 minutes. What time is it done?	24. Cook for 15 minutes. Stir. Then cook for 7½ minutes more. What is the total cooking time?
25. Lisa's casserole will cook for 45 minutes. To be done at 6:30, what time should it start cooking?	26. A grill holds 4 hamburgers. You need to cook 12 in all. How many batches of burgers must you cook?
27. A roast cooks for 25 minutes per pound. Emma's roast is 6 pounds. How long will it take to cook?	28. Bev buys cleaning supplies on sale for $5.79. The regular cost would be $7.99. How much does Bev save?
29. A set of 4 dinner napkins costs $9. How much for one napkin?	30. Barbecue tools cost $12 plus 5% sales tax. What amount is the tax?
31. Lou's coffeemaker cost $35. He got a discount of 10% off the price. What amount did he save?	32. Jay buys a new range for $420. He has 24 months to pay. How much will he pay each month?

Name _____   Date _____

Counting Calories

Calories in the foods you eat supply your body with energy. To stay in good shape, people need different amounts of calories each day. Look at this chart. How many calories should you eat every day?

Age	Girls	Boys
11–14 years old	2300 calories	2800 calories
15–18 years old	2300 calories	3000 calories
19–22 years old	2000 calories	3000 calories

Part 1. Add to find the total calories in each meal.

1. toast and butter 160
   fried egg 100
   tomato juice 25
   milk 165

   TOTAL calories: _____

2. plain yogurt 120
   banana 100
   black coffee 0

   TOTAL calories: _____

3. chicken salad 175
   corn muffin 155
   1 tomato 30
   skim milk 90
   tangerine 40

   TOTAL calories: _____

4. hamburger 370
   French fries 155
   milkshake 520

   TOTAL calories: _____

5. pork chop 130
   lima beans 75
   applesauce 90
   milk 165
   sherbet 120

   TOTAL calories: _____

6. chicken potpie 485
   green salad 110
   ginger ale 80
   baked apple 160

   TOTAL calories: _____

Part 2. Use your answers above to do these problems.

7. Which meal has the highest total calories? Meal # _____

8. Which meal has the lowest total calories? Meal # _____

9. Carla is 14 years old. On Tuesday, she ate meals #1 and #4 and #6.
   Did those meals provide enough calories for a girl her age? _____

10. Mark is 17 years old. He ate meals #2 and #3 and #5.
    For a boy his age, did Mark get too few or too many calories ? _____

Nutrition Facts

Many food packages have "Nutrition Facts" labels. These labels give you information about the food inside. You can choose nutritious foods for a healthy diet if you know the facts.

Directions. Look at this "Nutrition Facts" label. The label is from a box of oatmeal cereal. Use the facts on the label to answer the questions.

1. CALORIES. A serving of oatmeal cereal with skim milk has 190 calories. The cereal alone has 150 calories. How many calories come from the milk?

   _____

2. FAT. The total Percent Daily Value of fat for one day is 100 percent. A serving of oatmeal gives you 5 percent of the total. What percent of the day's fat will come from other foods?

   _____

3. IRON. The total Percent Daily Value of iron for one day is 100 percent. A serving of oatmeal gives you 10 percent of the total. What percent of the day's iron will come from other foods?

   _____

4. FIBER. A bowl of oatmeal with milk gives you 4 grams of dietary fiber. Suppose you need to get 30 grams in one day. How many grams will come from other foods?

   _____

5. CARBOHYDRATES. A bowl of oatmeal with milk gives you 27 grams of carbohydrates. Suppose you should get 300 grams in one day. How many grams will come from other foods?

   _____

Nutrition Facts

Serving Size: ½ cup dry (40 g)
Servings Per Container: 13

Amount Per Serving

	Cereal Alone	With ½ cup of Vit. A & D Fortified Skim Milk
Calories	150	190
Calories From Fat	25	25

	% Daily Value *	
Total Fat 3 g	5%	5%
Saturated Fat 0.5g	2%	2%
Polyunsaturated Fat 1g		
Monounsaturated Fat 1g		
Cholesterol 0mg	0%	0%
Sodium 0mg	0%	3%
Total Carb 27g	9%	11%
Dietary Fiber 4g	15%	15%
Sugars 1g		
Protein 5g		
Vitamin A	0%	4%
Vitamin C	0%	2%
Calcium	0%	15%
Iron	10%	10%

* Percent Daily Values are based on a 2,000 calorie diet. Your daily values may be higher or lower depending on your calorie needs:

	Calories	2,000	2,500
Total Fat	Less than	65g	80g
Sat Fat	Less than	20g	25g
Cholesterol	Less than	300mg	300mg
Sodium	Less than	2,400mg	2,400mg
Total Carbohydrate		300g	375g
Dietary Fiber		25g	30g

Calories per gram:
    Fat 9 • Carbohydrate 4 • Protein 4

Name _____   Date _____

Figuring the Cost of One Meal

Suppose your club gave you $25 to buy food for a holiday dinner to feed six people. You had to spend as much of the $25 on food as you could, WITHOUT GOING OVER $25.

Decide which of the meats below you might want for the dinner. Write the meat you chose and its price on line 1. Then select all the other foods you'd like to serve with the meat. Write those foods and their prices on the other lines. (The amounts on the list are correct for serving six people.)

TURKEY $12.50

HAM $11.90

ROAST BEEF $9.55

FOOD PRICE LIST

Lettuce	1.19	Broccoli	1.29
Carrots	.39	Potatoes	.55
Tomatoes	.98	Corn	1.98
Onion	.38	Green Beans	1.65
Olives	1.45	Stuffing	1.69
Salad Dressing	1.79	Cranberry Sauce	.95
Mustard	.79	Milk	2.16
Applesauce	2.19	Fruit Drink	1.78
Pickles	1.39	Pineapple Sherbet	2.99
Bread	1.29	Pumpkin Pie	3.49
Margarine	.89	Lemon Cookies	2.18

PRICE

1. The meat I chose: _____ $ _____

2. Other foods I chose: _____   _____

_____   _____

_____   _____

_____   _____

_____   _____

3. The total cost of food for the dinner: _____ $ _____

4. I had $ _____ left over from the $25.

Name _____   Date _____

The Food Budget

A budget is a plan for spending money. Many people plan how much money they want to spend for food each week. They try not to spend more than the amount in their budget.

Example: Mrs. Brown's food budget is $55 per week. So far, she has spent a total of $41.

Problem: How much does she have left for the rest of the week?

Solution:
Amount in the budget:	$55
Subtract the amount spent:	–41
She has left:	$14

Directions. Solve the budget problems below. You may need another sheet of paper to do the work.

---

1. Tony planned to spend $35 for food this week. On Monday he spent $22. How much was left?

    _____

2. Kathy's food budget was $30 per week. She spent $23 on Thursday. How much did she have for the rest of the week?

    _____

---

3. Richard spent $11 to buy a turkey. He spent $27 for other foods. His weekly food budget was $50.

    a. How much did he spend? _____
    b. How much is left? _____

---

4. This week, Brenda spent $10 and $14 and $22 for groceries. She planned a budget of $50 for food.

    a. How much did she spend? _____
    b. Was it more than the budget? _____

---

5. Kate and Jon try not to spend more than $40 per week for their groceries. This week, Jon spent $19 and Kate spent $27.

    a. How much did they spend? _____
    b. Was it more than the budget? _____

---

6. The O'Connor family budgets $500 per month for food. They spend $135 the first week, $123 the second week, and $112 the third week.

    a. How much have they spent so far?   _____
    b. How much is left for the fourth week?   _____

Kitchen Math

Name _____ Date _____

Planning a Cookout

The class wants to have a cookout. The students are planning the menu now. They need to know how much food to buy for everyone.

Example: There are 24 students. Each student will eat 3 cookies.
One package contains 16 cookies.

Problem: How many packages do they need to buy?

Solution: This is a two-step problem.

a. 24   students
     × 3   cookies each
     72   cookies in all

They will need 72 cookies.

b.
$$\begin{array}{r} 4 \text{ r.}8 \text{ packages needed} \\ 16 \text{ per package} \overline{)\ 72 \text{ cookies in all}} \\ -64 \\ \hline 8 \end{array}$$

They need to buy 5 packages.

Here are the foods for the cookout. Look at the quantities.

hot dogs (12 per pack)   fruit drink (32 oz per can)
buns (8 per bag)   watermelon (12 slices per melon)
tortilla chips (36 oz per bag)   cookies (16 per pack)

Directions. Find out how much of each food the class will need to buy. You may need another sheet of paper to do the work.

For each student	For 24 students?	How many to buy?
3 cookies ×24 =	_72_ cookies 16)72 4 −64 8	_5_ packages
2 hot dogs	____ hot dogs	____ packages
2 buns	____ buns	____ bags
6 oz tortilla chips	____ ounces	____ bags
8 oz fruit drink	____ ounces	____ cans
2 slices watermelon	____ slices	____ melons

Name _____ Date _____

Review on Planning Ahead

Directions. Do the problems on another sheet of paper. Choose the *best* operation to solve each problem. Circle its letter in the correct column. Then write it in the blank in the riddle. One has been done for you.

Problems	+	–	×	÷	Answers
1. An apple has 70 calories. A donut has 135. How many calories in both?	Ⓛ	Y	D	M	205
2. Eating 2 tablespoons of peanut butter gives you 200 calories. Suppose you should get 2300 calories in one day. How many calories will come from other foods?	I	U	T	G	
3. Twelve rolls are in a package. How many in three packages?	R	K	C	E	
4. Four people share a box of 20 cookies. How many cookies each?	A	C	W	S	
5. A jelly bean has seven calories. How many calories in six beans?	J	B	E	I	
6. The total Percent Daily Value of iron for one day is 100 percent. A serving of baked beans gives you 25 percent of the total. What percent of the day's iron will come from other foods?	M	L	R	F	
7. Six people are here. Each wants three eggs. How many do you need?	T	B	K	O	
8. A turkey costs $12.75. You have $10. How much more do you need?	W	L	P	V	
9. A box has 45 chicken nuggets. There are five people. How many will each person get?	Q	C	J	D	
10. Corn costs $.99 and butter costs $1.89. How much for both?	P	R	M	E	
11. You had $15. You spent $13.29 for food. How much is left?	Z	I	F	N	

Riddle: What do you serve at a boring party?

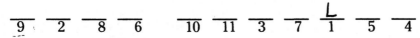

$$\overline{\ 9\ }\ \overline{\ 2\ }\ \overline{\ 8\ }\ \overline{\ 6\ }\quad \overline{10}\ \overline{11}\ \overline{\ 3\ }\ \overline{\ 7\ }\ \overset{L}{\overline{\ 1\ }}\ \overline{\ 5\ }\ \overline{\ 4\ }$$

11   *Kitchen Math*

The Shopping List

Smart shoppers make a list of what they need to buy. When you go shopping, take your list with you. It will help you to remember what you need. It can also help you to avoid buying things that you don't need.

BEST BUY MARKET

Orange Juice, 32 oz	.99	Bran Bread, loaf	2.29
Cheese Slices, 8 oz	1.49	Veg. Oil, 24 oz	1.79
Frozen Deluxe Pizza	4.29	Ice Cream, ½ gal	2.99
Parmesan Cheese, 12 oz	3.59	Large Eggs, doz	1.59
Thin Spaghetti, 1 lb	.67	Lo-Fat Milk, gal	1.99
Lettuce, head	.99	Lean Hamburger, lb	1.99

Directions. Using the prices above, find the total cost of the items on each person's shopping list. You may need another sheet of paper to do the problems. The first is done for you.

Example

		$ Each	Cost
2 pizza	2 x	4.29	8.58
1 lb hamburger		1.99	1.99
2 bread	2 x	2.29	4.58
2 milk	2 x	1.99	3.98
		TOTAL:	$ 19.13

1. BETH'S LIST

	$ Each	Cost
1 milk		
1 orange juice		
2 lettuce		
1 doz eggs		
	TOTAL:	

2. AMY'S LIST

	$ Each	Cost
1 oil		
1 lb hamburger		
2 lettuce		
1 lb spaghetti		
	TOTAL:	

3. JEFF'S LIST

	$ Each	Cost
2 spaghetti		
2 Parm cheese		
1 bread		
1 ice cream		
	TOTAL:	

4. JUAN'S LIST

	$ Each	Cost
1 orange juice		
1 pkg cheese slices		
3 bread		
2 doz eggs		
	TOTAL:	

5. LEE'S LIST

	$ Each	Cost
4 froz. pizza		
3 lo-fat milk		
2 lb hamburger		
2 ice cream		
	TOTAL:	

Count Your Change

Always count your change when you pay for your groceries. Even if the cash register is a computer, you should check the change a cashier gives you. The cashier could make a mistake.

Example: Your groceries cost $7.50. You give the cashier a $10.00 bill.

Problem: How much change should you receive?

Solution:

Amount you gave the cashier:	$10.00
Subtract the cost of the groceries:	– 7.50
Your change:	$2.50

Part 1.   How much change should you receive?

1. $7.00	2. $11.00	3. $5.50	4. $25.00	5. $13.01
– 6.00	– 10.50	– 5.20	– 24.97	– 12.01

Part 2.   Here is Ray's cash register slip from a supermarket. Look at Ray's slip to answer the questions below.

BEST FOOD STORE	
tuna fish	.65
dog biscuits	1.39
pancake mix	.89
margarine	.59
low-fat milk	2.06
frozen peas	.89
NONTAXABLE	$ 5.08
TAXABLE	$ 1.39
TAX	$ .07
TOTAL	$ 6.54
CASH TDR	$ 7.00
CHANGE DUE	$ .46

6.   What was the total cost of Ray's groceries?   _____

7.   How much money did Ray give to the cashier?   _____

8.   How much change should Ray have received?   _____

9.   The cashier gave Ray one quarter, two dimes, and one penny. Was that the correct change?   _____

10.   On what item did Ray have to pay tax?   _____

© 1988, 1997 J. Weston Walch, Publisher

Kitchen Math

Fresh Produce by the Pound

Try to buy fruits and vegetables "in season" when they are fresh. They probably will cost less and taste better. In the produce section of your store, fruits and vegetables often are priced by the pound. To find out how much they will cost, you must multiply the price per pound times the weight.

_____

Directions. Find the cost of the food in each problem below. The first one has been done for you.

Example:

$ .33
× 3
$ .99

3 pounds
$.333/lb.
Answer: _____

1. CARROTS

2 pounds
$.45/lb.
Answer: _____

2. ORANGES

3 pounds
$.79/lb.
Answer: _____

3. POTATOES

5 pounds
$.39/lb.
Answer: _____

4. ONIONS

5 pounds
$.49/lb.
Answer: _____

5. APPLES

6 pounds
$.99/lb.
Answer: _____

6. MUSHROOMS

2 pounds
$2.69/lb.
Answer: _____

7. GRAPES

3 pounds
$1.49/lb.
Answer: _____

8. Connie picks out 6 apples. These apples are priced to sell at $1.50 per pound. The apples weigh 1.5 pounds. What will Connie pay? _____

9. A sign says that red potatoes are $.70/lb. If Lin buys 4.5 pounds, how much will the potatoes cost? _____

10. Marco buys some tomatoes. They weigh 2.6 pounds. At $1.30 per pound, what will Marco pay? _____

Read the Label

Many meats and other foods are priced by the pound. To find out the price, you must understand how to read the labels. Look at this label from a package of meat.

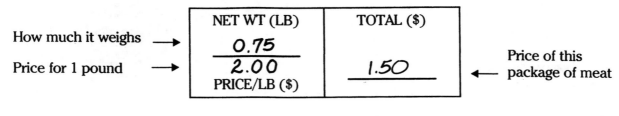

How much it weighs ⟶

Price for 1 pound ⟶

NET WT (LB)	TOTAL ($)
0.75	
2.00	1.50
PRICE/LB ($)	

⟵ Price of this package of meat

Multiply the price per pound times the weight of the food.

$2.00 PER LB
× .75 LB
$1.50 TOTAL

Directions. Read the labels below. Find out the price of each package. You may need another sheet of paper to do the problems. The first one is done for you.

Ex.

NET WT (LB)	TOTAL ($)
0.50	
1.28	# .64
PRICE/LB ($)	

$1.28
× .50
$.6400

1.

NET WT (LB)	TOTAL ($)
1.00	
.79	
PRICE/LB ($)	

2.

NET WT (LB)	TOTAL ($)
3.00	
.44	
PRICE/LB ($)	

3.

NET WT (LB)	TOTAL ($)
0.50	
1.98	
PRICE/LB ($)	

4.

NET WT (LB)	TOTAL ($)
1.50	
2.18	
PRICE/LB ($)	

5.

NET WT (LB)	TOTAL ($)
3.20	
3.05	
PRICE/LB ($)	

Name _____   Date _____

How Much for One?

Pat saw this price marked on a carton of yogurt. It seemed like a good price. But Pat wanted to buy only one carton.

Problem: How much would one carton cost?

Solution: Divide the price by the quantity.

$$\begin{array}{r} \$\,0.49 \text{ r.} 2 \\ 3\overline{)\$1.49} \\ \underline{-12} \\ 29 \\ \underline{-27} \\ 2 \end{array}$$

The problem does not come out even. The price for one carton will be more than $.49, so the store will round the amount up to the next penny. Pat will pay $.50 for one carton.

Part 1. How much will you pay for one?

1. 5/$1.00 _____

2. 5/$.99 _____

3. 3/$.75 _____

4. 3/$.79 _____

5. 4/$1.29 _____

6. 4/$1.49 _____

Part 2. How much will they pay?

7. Randy wants a can of frozen juice. The price is marked 3/$1.00 on the can. What will one can cost? _____

8. Kim sees a jar of mustard. It is marked 2/$.95. How much will the store charge Kim for one jar? _____

9. A 26-ounce carton of salt is marked 2/$.89 but Jessica needs only one carton. What will the price be for one? _____

10. A package of buttermilk biscuits was priced at 4/$1.19. Julie wants to buy only 2 packages. How much will they cost? _____

16   *Kitchen Math*

Buying Baby Food

What kinds of food does a baby need? How much baby food should you buy? How much will it cost?

These are important things to know when you shop for baby food. Babies and little children eat small amounts of food. But they eat more times each day than adults. Baby food can make up a big part of a family's food budget.

Part One. How long will it last? Solve the baby food problems below. You may need another sheet of paper to do the work.

1. Marta's baby eats 3 servings of baby cereal each day. An 8-oz box of cereal makes 15 servings. How many days will one box of cereal last? _____

2. Sarah's baby has 2 servings of baby cereal each day. A 16-oz box of cereal makes 30 servings. How many days will one box of cereal last? _____

3. Pat's baby takes 2 bottles of juice each day. A small baby bottle holds 4 ounces. A jar of juice contains 32 ounces.

   a. How many baby bottles will one jar of juice fill? _____

   b. How many days will one jar of juice last? _____

4. Katy's baby takes 4 bottles of formula each day. A baby bottle holds 8 ounces. After mixing with water, a can of formula makes 112 ounces.

   a. How many baby bottles will one can of formula fill? _____

   b. How many days will one can of formula last? _____

Part Two. How much will it cost?

5. Terry's son eats 3 medium jars of baby food each day. One jar costs 52¢. How much will the baby food cost:

   a. For one day? _____

   b. For 7 days? _____

6. Ben's baby eats 5 small jars of baby food each day. One jar costs 37¢. How much will the baby food cost:

   a. For one day? _____

   b. For 7 days? _____

Review on Shopping for Food

Directions. Do the following problems on another sheet of paper. Choose the *best* operation to solve each problem Circle its letter in the correct column. Then write it in the blank in the riddle. One has been done for you.

Problems	+	−	×	÷	Answers
1. Bread costs $.60 and milk costs $1. How much for both?	Ⓐ	L	D	N	#1.60
2. Roast beef is $3.99 a pound. How much for five pounds?	B	T	E	U	
3. A box of baby cereal is $1.09. How much do three boxes cost?	P	K	S	I	
4. Grapefruits are priced at 4/$1. How much does one cost?	A	C	Y	O	
5. Carrots cost $.99 a bunch. What will four bunches cost?	J	L	O	S	
6. Your groceries cost $27.50. How much change do you get from $30?	V	P	R	F	
7. A package of chicken is marked $.80/lb. Net weight is 1.50 lbs. What is the total cost?	E	I	T	O	
8. A baby bottle holds 8 ounces. A can of ready-to-use formula contains 32 ounces. How many baby bottles will one can of formula fill?	W	L	S	T	
9. The total cost of your groceries is $12.89. You pay with a $20 bill. How much is your change?	C	T	R	D	
10. Cans of juice are marked 3/$1. What will you pay for one can?	N	T	M	A	
11. Apples are $.80 per pound. How much for 5.6 pounds?	O	I	M	N	

Riddle: What did Goofy Gus use for glue?

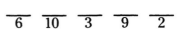
__ __ __ A __ __   __ __ __ __ __
7   4 11 1 8 5   6 10 3 9 2

Kitchen Math

Name _____   Date _____

Saving Money with Coupons

Coupons printed in newspapers and magazines can save you money on food. If you want to buy the item shown on the coupon, cut out the coupon and take it to the store. Give your coupon to the cashier when you pay for the item.

Example: This coupon is worth 20 cents off the price of a box of cereal.

Problem: How much will you pay for the cereal if you use the coupon at the store?

Solution: Find out the price of the cereal: $1.40
Subtract the worth of the coupon: −.20
You will pay: $1.20

Part 1.   How much will you pay?

1. $1.50
     − .50

2. $2.35
     − .25

3. $.99
     − .15

4. $.79
     − .12

5. $1.05
     − .15

Part 2.   This coupon was printed in the newspaper. Look at the coupon and answer the questions below.

6.   What amount of money can this coupon save you?   _____

7.   Does the coupon tell you the price of the item?   _____

8.   Can you use the coupon any time you want?   _____

9.   If the regular price is $.95, what will you pay if you use the coupon?   _____

10.   If the regular price is $1.00, what will you pay if you use the coupon?   _____

Kitchen Math

Comparing Store Prices

Food is sold at small convenience stores, medium-sized grocery stores, and large supermarkets. Smart shoppers check at several stores to find the best prices. Use the prices shown below to answer the questions. You may need another sheet of paper to work the problems.

ITEM	Corner Variety	Elm Street Market	Mammoth Foods
LETTUCE	$1.25/head	$ .99/head	$ .79/head
TUNA	$1.14/can	$ .99/can	$1.03/can
MARGARINE	$1.29/pkg.	$ .99/pkg.	$ .55/pkg.
KETCHUP	$1.40/bottle	$1.29/bottle	$1.29/bottle
APPLES	$1.39/bag	$1.29/bag	$1.35/bag
RAISINS	$1.59/box	$1.49/box	$1.38/box

Example: LETTUCE

a. What is the highest price? $1.25

b. What is the lowest price? − .79

c. Find the difference: $ .46

1. TUNA

a. What is the highest price? _____

b. What is the lowest price? _____

c. Find the difference: _____

2. MARGARINE

a. What is the highest price? _____

b. What is the lowest price? _____

c. Find the difference: _____

3. KETCHUP

a. What is the highest price? _____

b. What is the lowest price? _____

c. Find the difference: _____

4. APPLES

a. What is the highest price? _____

b. What is the lowest price? _____

c. Find the difference: _____

5. RAISINS

a. What is the highest price? _____

b. What is the lowest price? _____

c. Find the difference: _____

6. Nancy needs all of the items listed on this page. Write the total amount Nancy would pay:

   a. At Corner Variety _____

   b. At Elm Street Market _____

   c. At Mammoth Foods _____

7. Where should Nancy buy her food? _____

8. How much could she save by shopping at the least expensive store instead of at the most expensive? _____

Kitchen Math

Comparing Brands

When you shop for food, check the brand names. Compare the price of one item sold by different brands. Usually, a major-name brand will cost more than a store's own brand. Foods with a generic (or "no-name") brand may have the lowest prices of all.

Directions. Use the canned food prices below to answer the questions. You may need another sheet of paper to do the work.

	MAJOR BRAND	SHOP-RITE	GENERIC
Peaches	$1.15	$1.09	$.99
Wax Beans	$.79	$.53	$.49
Pears	$1.15	$1.09	$1.05

1. WAX BEANS

a. Which brand has the highest price? _____   What is that price? _____

b. Which brand has the lowest price? _____   What is that price? _____

c. What is the difference between the lowest and highest?   _____

d. How much could you save if you got 2 cans? _____

2. PEACHES

a. Which brand has the highest price? _____   What is that price? _____

b. Which brand has the lowest price? _____   What is that price? _____

c. What is the difference between the lowest and highest?   _____

d. How much could you save if you got 4 cans? _____

3. PEARS

a. Which brand has the highest price? _____   What is that price? _____

b. Which brand has the lowest price? _____   What is that price? _____

c. What is the difference between the lowest and highest?   _____

d. How much could you save if you got 3 cans? _____

Unit Pricing

Which jar of peanut butter is the better buy? The retail price is higher for the 28-ounce jar of Goodie. But compare the unit prices.

a.   b.

UNIT PRICE	RETAIL PRICE
11.0¢ PER OZ	**3**.08
	28 OZ
GOODIE PEANUT BUTTER	

UNIT PRICE	RETAIL PRICE
12.0¢ PER OZ	**2**.16
	18 OZ
YUMMIE PEANUT BUTTER	

Some stores do not have signs like these. You must figure out the unit prices by yourself.

To find the unit price, divide the units into the retail price.

a. GOODIE $.11 per oz
28 oz) $3.08 retail

b. YUMMIE $.12 per oz
18 oz) $2.16 retail

Directions. Find the unit prices for each item below. You will need another sheet of paper to work the problems. Which is the better buy—(a) or (b)?

Find the unit prices for each item below. You will need another sheet of paper to work the problems. Which is the better buy—(a) or (b)?

Item	Retail Prices		Unit Prices	Better Buy
1. Yogurt	a.	3 cartons for $1.95	_____	
	b.	2 cartons for $1.45	_____	_____
2. Ginger Ale	a.	6-pack for $1.68	_____	
	b.	8-pack for $2.12	_____	_____
3. Flour	a.	5 lbs for $1.55	_____	
	b.	2 lbs for $1.05	_____	_____
4. Macaroni	a.	3 lbs for $1.68	_____	
	b.	2 lbs for $1.09	_____	_____
5. Apple Juice	a.	$2.88 for 64 oz	_____	
	b.	$1.76 for 32 oz	_____	_____

Buying Large Quantities

One way to save money is to buy foods in large amounts. But you should be able to use up the food you buy. And you must know how to compare the quantities and costs.

Example: Jake's Warehouse Club sells macaroni in 3-lb boxes for $1.17.
Another store sells smaller 1-lb boxes for $.59.

Problem: How much can you save if you buy 3 pounds at Jake's?

Solution: a. Divide to compare quantities:
Three 1-lb boxes make 3 pounds.

$$\overset{\text{3 small boxes}}{1 \text{ lb } \overline{)3 \text{ lb}}}$$

b. Multiply to find the cost:
Three 1-lb boxes cost $1.77.

$.59 \times 3$ small boxes = $1.77 for 3 lb

c. Subtract to find the savings:
You can save $.60 at Jake's.

$1.77 - $1.17 at Jake's = $.60 savings

Directions. Compare the quantities and costs to solve the problems. You may need another sheet of paper to do the work.

1. Rice comes in 20-lb bags for $6.49 at Jake's. Another store sells rice in smaller amounts.

Compare:		2-lb bag — $.99	5-lb bag — $2.39	10-lb bag — $3.89
a.	How many bags make 20 lb?			
b.	What is the cost to buy 20 lb?			
c.	What can you save at Jake's?			

2. At Jake's a 10-lb bag of sugar costs $3.79. Another store sells 5-lb bags for $2.09.

   a. What is the cost of 10 pounds if you buy the 5-lb bags? _____

   b. How much can you save if you buy the 10-lb bag at Jake's? _____

3. Flour comes in 25-lb bags for $5.99 at Jake's. Another store sells 5-lb bags for $1.49.

   a. What is the cost of 25 pounds if you buy the 5-lb bags? _____

   b. How much can you save if you buy the 25-lb bag at Jake's? _____

Name _____ Date _____

Review on Stretching Food Dollars

Directions. Do the problems on another sheet of paper. Choose the *best* operation to solve each problem. Circle its letter in the correct column. Then write it in the blank in the riddle. One has been done for you.

Problems	+	−	×	÷	Answers
1. You use a coupon for 25¢ off the price of a cake. The cake's price is $2.10. What will you pay?	O	Ⓕ	I	V	# 1.85
2. One brand of pears is $.89 and another brand is $.93. What is the difference in their prices?	J	L	C	F	
3. A bag of eight candy bars is $1.44. What is the price per bar?	P	D	O	S	
4. Rice comes in 20-pound bags and 5-pound bags. How many 5-pound bags make 20 pounds?	T	G	A	T	
5. One brand of rice costs 9¢ less per pound than another brand. How much do you save if you buy 15 pounds?	F	B	W	S	
6. A 1-pound box of baking soda costs $.59. What is the cost to buy 4 pounds?	L	H	I	Y	
7. At Smith's Store, frozen pizza is $3.15. At Shop-Mart it is $2.59. What can you save at Shop-Mart?	E	O	R	X	
8. Which is the better buy: 3 cans for $.84 or 4 cans for $1.00?	M	C	G	F	
9. A 3-liter bottle of cola costs $1.79. A coupon saves 20¢. What do you pay?	A	E	T	S	
10. A 5-lb bag of flour is 31¢ per pound. A 2-lb bag is 47.5¢ per pound. What is the difference in prices per pound?	R	S	N	I	
11. A 32-ounce bottle of orange juice is 96¢. What is the cost per ounce?	U	W	K	B	

Riddle: What was wrong with the macaroni?

$$\overline{10}\ \overline{4}\ \overline{6}\ \overset{F}{\overline{1}}\ \overline{8}\qquad \overline{9}\ \overline{2}\ \overline{11}\ \overline{7}\ \overline{5}\ \overline{3}$$

24

Kitchen Math

Name _____   Date _____

Equal Measures

Success in cooking begins with correct measuring. To get the best results, keep in mind these basic equivalents.

3 teaspoons	=	1 tablespoon		
16 tablespoons	=	1 cup		
1 cup	=	8 ounces	=	½ pint
2 cups	=	16 ounces	=	1 pint
4 cups	=	32 ounces	=	1 quart
2 quarts	=	64 ounces	=	½ gallon
4 quarts	=	128 ounces	=	1 gallon
1 square of chocolate		=	1 ounce	
2 tablespoons butter		=	1 ounce	
1 stick of butter		=	4 ounces	

Directions. Using the chart above, multiply to answer these questions.

1. How many squares would give you 4 ounces of chocolate?

   _____

2. Two pints of cream would give you how many ounces?

   _____

3. How many teaspoons would give you 3 tablespoons of parsley?

   _____

4. How many tablespoons would give you 4 ounces of butter?

   _____

5. Three quarts would give you how many cups of soup?

   _____

6. Four sticks of butter would give you how many ounces?

   _____

7. How many teaspoons of honey would give you 2 tablespoons?

   _____

8. How many cups would give you 2 quarts of water?

   _____

9. Two gallons of cider would give you how many quarts?

   _____

10. Sixteen tablespoons would give you how many ounces of juice?

    _____

Kitchen Math

Name _____ Date _____

Combining Liquid Ingredients

Serena is making fruit punch. She combines 1000 mL (1 liter) of apple juice, 500 mL of lemonade, and 1000 mL of lime soft drink. How much punch will Serena have?

To find out, add together all the ingredients.

$$
\begin{array}{rl}
1000 \text{ mL} & \text{juice} \\
500 \text{ mL} & \text{lemonade} \\
+ \quad 1000 \text{ mL} & \text{soft drink} \\
\hline
2500 \text{ mL} & \text{(total)}
\end{array}
$$

She will have 2500 mL (or 2.5 liters) of punch.

_____

Part 1. Add to find the total amounts.

1. 250 mL + 500 mL = _____

2. 140 mL + 160 mL = _____

3. 500 mL + 500 mL = _____

4. 250 mL + 700 mL = _____

5. 500 mL + 1000 mL = _____

6. 1.5 L + 2.5 L = _____

Part 2. How much will they have in all?

7. Mark combines two bottles of cola. One bottle contains 350 mL and the other contains 1000 mL. What is the total amount of cola?

   _____

8. In making soup, Patty adds 500 mL of tomato juice and 125 mL of water. How much liquid does she add in all?

   _____

9. Ellen's recipe includes 250 mL of cream, 750 mL of milk, and 500 mL of water. What is the total amount?

   _____

10. Tracy mixes 2 liters of ginger ale, 2 liters of lemon drink, and 1.5 liters of cranberry juice. How much liquid will Tracy have in all?

   _____

 Kitchen Math

Name _____   Date _____

Combining Dry Ingredients

Jared mixes 2½ cups of rye flour and 2½ cups of white flour. How many cups will he have in all?

Let's add to find out.

> 2½ cups
> + 2½ cups
> 4² cups = 5 cups in all

Part 1. Add to find the total amounts.

1.   2   cups
   + 1½   cups

2.   3   cups
   +   ¼ cup

3.   1⅔ cups
   + 1   cup

4.   2½   cups
   + 2   cups

5.   ⅓ cup
   + ⅔ cup

6.   2¼ cups
   + ½ cup

Part 2. How much will they have in all? You may need another sheet of paper to do the work.

7.   Tina uses ½ cup of brown sugar and ½ cup of granulated sugar. How much sugar in all?

8.   Lan combines 2½ cups of flour and 2 cups of rolled oats. How many cups is that?

9.   Victor mixes 1¼ cups of ziti with 1¼ cups of rotini. How many cups in all?

10.   Senta has 1¾ cups of white flour. She adds ¾ cup of rye flour. How much flour in all?   _____

Kitchen Math

Name _____   Date _____

Figuring Leftovers

Betty Ann has 5 pounds of beef. She will use 3½ pounds to make stew. How many pounds of beef will she have left over?

Let's subtract to find out.

$$\begin{array}{ll} \cancel{5}\ ^{4}\ \cancel{3/2} & \text{lb to start} \\ -\ 3\frac{1}{2} & \text{lb for stew} \\ \hline 1\frac{1}{2} & \text{lb left over} \end{array}$$

Part 1. Subtract to find out how much is left.

1.	3	lb	2.	2	lb	3.	6	lb
	– 2½	lb		– ¼	lb		– 1½	lb

4.	5	cups	5.	3½	cups	6.	2½	cups
	– 4½	cups		– 2½	cups		– ¼	cup

Part 2. How much will they have left? You may need another sheet of paper to do the work.

7.   Anita has 3 cups of brown sugar. She uses ½ cup
     to make peanut butter cookies. How much is left?   _____

8.   Marlene has a pound of butter. She uses ¼ pound
     to make icing. How much butter is left?   _____

9.   Andy got 5½ pounds of hamburger. He cooks up
     2½ pounds of it. How much is left over?   _____

10. Bobbi has 1¼ cups of potato salad. She eats
     ¾ cup of it. How much potato salad is left?   _____

     Kitchen Math

Adjusting a Recipe

Recipes usually tell you how many servings they will yield. But sometimes you will want to make a different number of servings. When that happens, you must adjust the amount of each ingredient in the redcipe. Look at this chart.

HALF RECIPE $\div 2$	To get half as many servings, you make half a recipe. Divide the amount of each ingredient by 2.
DOUBLE RECIPE $\times 2$	To get twice as many servings, you double the recipe. Multiply the amount of each ingredient by 2.

Directions. Here is the list of ingredients from a recipe that serves four people. Fill in the chart to adjust the quantities for making a half recipe and a double recipe.

ITALIAN MEATBALLS to serve 4	HALF RECIPE to serve 2	DOUBLE RECIPE to serve 8
3 slices dry bread	_____	_____
$1\frac{1}{2}$ pounds ground beef	_____	_____
2 eggs	_____	_____
$\frac{1}{2}$ cup grated Romano cheese	_____	_____
1 tablespoon butter	_____	_____
4 tablespoons chopped onion	_____	_____
2 tablespoons chopped parsley	_____	_____
1 clove garlic, minced	_____	_____
$\frac{1}{2}$ teaspoon crushed oregano	_____	_____
$\frac{1}{4}$ teaspoon salt	_____	_____

Name _____   Date _____

Ratios in the Kitchen

A ratio is a way of comparing two amounts. In the kitchen, ratios are often used to figure out "before and after" amounts.

Example: One cup of uncooked rice will make 3 cups of cooked rice.

The ratio of uncooked to cooked is 1 to 3.
The fraction $\frac{1}{3}$ names this ratio

Problem: You want to make 6 cups of cooked rice. How many cups of uncooked rice should you use?

Solution: For rice, the ratio of "before" to "after" is $\frac{1}{3}$.

You know you want 6 cups "after." The equivalent ratio is:

$$\text{before} \rightarrow \quad \frac{1}{3} = \frac{?}{6} \quad \leftarrow \text{before}$$
$$\text{after} \rightarrow \qquad\qquad\qquad \leftarrow \text{after}$$

Now make the fractions equivalent. To get 6 cups of cooked rice, you need 2 cups of uncooked rice.

$$\frac{1}{3} = \frac{2}{6}$$

Part 1. Make the fractions equivalent.

1. $\frac{1}{2} = \frac{?}{8}$ _____

2. $\frac{3}{4} = \frac{?}{8}$ _____

3. $\frac{2}{1} = \frac{?}{3}$ _____

4. $\frac{1}{5} = \frac{2}{?}$ _____

5. $\frac{3}{1} = \frac{9}{?}$ _____

6. $\frac{2}{2} = \frac{6}{?}$ _____

Part 2. Figure the answers to these before-and-after problems.

7. If you squeeze one orange, you will get 4 ounces of juice. You want to make 32 ounces of juice. How many oranges must you squeeze? _____

8. Four ounces of cheese makes one cup of shredded cheese. You want 3 cups of shredded cheese. How many ounces must you shred? _____

9. Two slices of bread will make a cup of bread crumbs. If you want to make 3 cups of crumbs, how many slices of bread will you need? _____

10. If one pound of dry spaghetti makes 5 cups cooked, how many cups will 2 pounds make? _____

Kitchen Math

Review on Using Measurements

Directions. Do the problems on another sheet of paper. Choose the *best* operation to solve each problem. Circle its letter in the correct column. Then write it in the blank in the riddle. One has been done for you.

Problems	+	−	×	÷	Answers
1. You mix 250 mL of water and 650 mL of water. How much water do you have?	Ⓞ Y	C	B		*900 ml*
2. There are 8 ounces in a cup. How many ounces in 6 cups?	Z	S	A	E	
3. You have 4 pounds of meat. You cook $2\frac{1}{2}$ pounds. How much is left?	P	N	C	O	
4. A recipe calls for 3 cups of milk. How many cups for a half recipe?	W	D	O	S	
5. Mix $\frac{1}{4}$ cup of instant cocoa and $\frac{1}{4}$ cup of instant coffee. How much mixture in all?	E	I	A	D	
6. Combine 1 liter of soda, 1.5 liters of juice, 2 liters of cider. How much liquid in all?	U	E	N	R	
7. Mix $\frac{1}{2}$ cup sugar and $1\frac{1}{4}$ cups flour. How many cups in all?	A	L	S	F	
8. There are 32 ounces in a quart. How many ounces in 5 quarts?	T	G	O	V	
9. A recipe that makes 4 servings has 3 pounds of beef. How many pounds for making 8 servings?	P	W	R	M	
10. You have $3\frac{1}{2}$ cups of sugar. You use $1\frac{1}{4}$ cups. How many cups are left?	L	F	I	V	
11. Two celery stalks make 1 cup chopped. How many cups would 6 stalks make?	Y	X	L	E	

Riddle: What does a pound of rye bread do on a holiday?

$$\overline{}_{11}\ \overline{}_{8}\ \overline{}_{2}\ \overline{}_{10}\ \overline{}_{4}\qquad \overline{}_{7}\ \overline{}_{9}\ \overline{O}_{1}\ \overline{}_{6}\ \overline{}_{3}\ \overline{}_{5}$$

Kitchen Math

Name _____ Date _____

Cooking Times

Ron does not have a kitchen timer. How can he tell when his food will be done? He must look at the clock.

Example: Ron wants to cook frozen peas for 5 minutes.
He starts at 3:30.

Problem: What time will the peas be done?

Solution: Add 5 minutes to the time they started to cook.
The minute hand on the clock will have moved
ahead 5 minutes.

$$3:30 \;+\; 5 \text{ minutes} \;=\; 3:35$$

Part 1.   What time is it? Write the correct time for each clock.

1. _____   2. _____   3. _____

Part 2.   What time will it be in 15 minutes?

4. 10:00 _____   5. 4:15 _____   6. 9:55 _____

Part 3.   What time will the food be ready?

7. Jan puts a pan of brownies in the oven at
2:00. They bake for 30 minutes.

READY AT _____

8. Hard-boiled eggs cook for 15 minutes.
Carlo starts to cook them at 6:20.

READY AT _____

9. Judy is making baked potatoes. They need
to cook for one hour. She starts cooking at
5:45.

READY AT _____

10. Jim starts to bake a banana bread at 8:10
and it will take 55 minutes.

READY AT _____

Microwave Cooking Times

On a microwave oven, you set the timer for how long you want to cook.

The timer on a microwave is a digital clock. The timer divides a minute into 60 seconds. And it divides an hour into 60 minutes. That is how you must set the cooking times. Look at the times on this chart.

Cooking Times in Minutes and Seconds		
TIME	**MEANS**	**PRESS**
¼ min .... 15 seconds		1 5
½ min .... 30 seconds		3 0
¾ min .... 45 seconds		4 5
1 min		1 0 0
1½ min		1 3 0
1 hour .... 60 minutes		6 0 0 0

Part 1. Write the numbers to press for each cooking time.

1. For 30 seconds.   Press: _____

2. For 3 min.   Press: _____

3. For 1¾ min.   Press: _____

4. For 12½ min.   Press: _____

Part 2. Find the total time in each microwave recipe.

5. HIGH for 5 min.
   Then MEDIUM for 17 min.
   Total: _____

6. HIGH for 15 min. Stir.
   Then HIGH for 8½ min.
   Total: _____

7. HIGH for 12½ min.
   Let stand for 10 min.
   Total: _____

8. MEDIUM for 60 sec.
   Let stand for 3 min.
   HIGH for 4 min.
   Total: _____

9. DEFROST for 10 min.
   HOLD time 5 min.
   HIGH for 23 min.
   Total: _____

10. DEFROST 3 lb of beef for 11 min/lb
    Let stand for 15 min.
    Total: _____

Kitchen Math

Serving a Meal on Time

Helen wants to serve pizza at 6:00 sharp. The pizza will take 45 minutes to prepare. What time should Helen begin making the pizza?

To find the "start" time, subtract the preparation time from the "finish" time.

Helen figures backwards 45 minutes from the "finish" time of 6:00 and gets a "start" time of 5:15.

Directions. Figure backwards to find the "start" time for each meal below.

1. Finish time   7:00
   Prep. time   60 minutes

   Start time?   _____

2. Finish time   5:30
   Prep. time   30 minutes

   Start time?   _____

3. Finish time   6:15
   Prep. time   2 hours

   Start time?   _____

4. Finish time   7:00
   Prep. time $1\frac{1}{2}$ hours

   Start time?   _____

5. Finish time   12:00
   Prep. time   25 minutes

   Start time?   _____

6. Finish time   1:30
   Prep. time   45 minutes

   Start time?   _____

7. Fran needs to allow $3\frac{1}{2}$ hours for preparing stew. She wants to serve the stew at 8:30. What time should she start to make the stew?   _____

8. Laura is making a casserole to serve at 7:00. The casserole takes 90 minutes to prepare. What time should Laura begin?   _____

9. Andrea's luncheon will be at 12:30. She figures it will take 25 minutes to prepare. When should she start?   _____

10. Billy is making dessert for the team. The dessert will take 50 minutes to fix. Billy wants to be ready by 7:00. When should he begin making the dessert?   _____

Name _____   Date _____

Cooking in Batches

Some foods cannot be cooked all at once. You must cook them in several batches.

Example: A recipe makes 72 cookies.   You can bake only 24 cookies at a time.
The cookies will bake for 8 minutes.

Problems: a. How many batches must you make?
b. How many minutes will it take to bake all of the cookies?

Solutions:

a.   3 batches
24 per batch)‾72 cookies

b.   8 minutes
× 3 batches
24 minutes

You must bake 3 separate batches.   They will take 24 minutes in all.

Directions. For each problem below, find out how many batches are needed. Then figure how long it would take to cook them.

		BATCHES	COOKING TIME
1.	You want 8 pieces of toast. Your toaster makes 2 pieces at at time. Toast takes 3 minutes.		
2.	A recipe makes 12 corn sticks. You can make only 4 sticks at a time. They bake for 20 minutes.		
3.	Biscuits bake for 12 minutes. The recipe makes 32 biscuits. You can make only 16 biscuits at a time.		
4.	Your griddle will hold 10 little pancakes. The recipe makes 50. They cook for 4 minutes.		
5.	A recipe makes 36 rolls. Your cookie sheet will hold 24 rolls. The rolls must bake for 15 minutes.		

How Long to Cook Meat?

Recipes do not always tell you the exact amount of time to cook meat. Some cookbooks have charts instead. This chart shows the approximate times for cooking roast beef just the way you want it.

BEEF—Standing rib roast, at 300°F:	
	Min per pound
Rare	20
Med. Rare	22
Medium	25
Well-done	28

Paula has a 6-pound roast. She wants to serve it medium. Use the chart to see how many minutes per pound it should cook.

a. Find the total minutes.

$$
\begin{array}{r}
25 \text{ min/lb} \\
\times\ 6 \text{ lbs} \\
\hline
150 \text{ min total}
\end{array}
$$

b. Change minutes to hours.

$$
\begin{array}{r}
2 \text{ hr } 30 \text{ min} \\
60\,\overline{)\ 150 \text{ min}} \\
-120 \\
\hline
30
\end{array}
$$

Paula's roast should cook for 2 hours and 30 minutes.

Directions. Use the chart above to figure out the apporoximate cooking time for each roast. You may need another sheet of paper to work the problems.

	HOW COOKED	POUNDS x	MIN PER LB =	TOTAL MIN =	HOURS?
1.	Rare	6			
2.	Rare	5			
3.	Medium	7			
4.	Well-done	8			
5.	Med. Rare	4			

Name _____   Date _____

Review on Preparation Time

Directions. Do the problems on another sheet of paper. Choose the *best* operation to solve each problem. Circle its letter in the correct column. Then write it in the blank in the riddle. One has been done for you.

Problems	+	−	×	÷	Answers
1. A roast will cook for 150 minutes. How long is that in hours?	U	S	N	Ⓔ	$2\frac{1}{2}$ hours
2. Nan's bread must bake for 60 minutes. If it starts at 2:15, when is it ready?	O	D	J	A	
3. A muffin pan holds 8 muffins. If a recipe makes 24 muffins, how many batches must you cook?	C	Y	V	E	
4. A microwave recipe says to cook ham for 15 minutes per pound. How many minutes should a 5-pound ham cook?	W	L	N	R	
5. Tom makes three batches of rolls. Rolls bake for 15 minutes. How long will it take in all?	S	E	A	I	
6. Lunch takes 20 minutes to prepare. It will be served at 12:30. What time should you start making lunch?	B	P	M	D	
7. A roast weighs 4 pounds. If you cook it 22 minutes per pound, how many minutes will it take?	H	K	C	F	
8. Ginger's dinner will take $1\frac{1}{2}$ hours to prepare. If she starts at 4:30, when will dinner be ready?	G	N	E	A	
9. A microwave recipe says to cook new potatoes at HIGH for $8\frac{1}{2}$ minutes. Then let stand for 5 minutes. How long will it take in all?	S	R	L	Y	
10. Pizza cooks for 25 minutes. To be done by 7:00, what time should it start cooking?	Z	K	U	E	

RIDDLE: What did Goofy Gus use for washing his car?

$$\overline{}_{9} \quad \overline{}_{6} \quad \overline{}_{2} \quad \overline{}_{4} \quad \overline{}_{8} \quad \overline{E}_{1} \qquad \overline{}_{7} \quad \overline{}_{5} \quad \overline{}_{10} \quad \overline{}_{3}$$

© 1988, 1997 J. Weston Walch, Publisher   37   *Kitchen Math*

Buying Cleanup Supplies on Sale

SAVE!	REG.	SALE
Spray cleaner	$1.79	$1.49
Floor wax	3.29	2.99
Paper towels	1.59	1.37
Sponges	1.25	.99
Dish soap	2.19	1.89
Cleanser	.73	.69
Pot scrubber	.89	.78
Oven cleaner	3.19	2.98
Trash bags	2.37	1.99

Ann is planning to have a party. She needs to get some supplies for cleaning up the kitchen after the party.

This ad was in the newspaper. The special sale prices will help Ann to save some money.

Directions. Using this ad, compare the regular prices and sale prices. You may need another sheet of paper to do the problems. How much can each person save at the sale prices?

1. ANN'S LIST

	Reg.	Sale
paper towels		
dish soap		
trash bags		
TOTALS:		

How much can Ann save? _____

2. TOM'S LIST

	Reg.	Sale
floor wax		
sponges		
pot scrubber		
TOTALS:		

How much can Tom save? _____

3. JOHN'S LIST

	Reg.	Sale
spray cleaner		
oven cleaner		
paper towels		
TOTALS:		

How much can John save? _____

4. LIZ'S LIST

	Reg.	Sale
dish soap		
cleanser		
sponges		
TOTALS:		

How much can Liz save? _____

Name _____   Date _____

Buying Kitchen and Table Linens

Kitchen towels help keep your hands and dishes neat and clean. Nice placemats, napkins, and tablecloths make your meals look more interesting.

When you buy kitchen and table linens, shop for good quality and good prices. Compare the costs. Should you buy the items separately, or should you buy a package?

Part One. Divide to find the cost for one item in each package.

1. Napkins — 4 for $6.96 _____   2. Napkins — 2 for $2.96 _____

3. Placemats — 6 for $22.50 _____   4. Placemats — 4 for $19.80 _____

5. Towels — 2 for $2.98 _____   6. Towels — 4 for $7.16 _____

Part Two. Multiply to find the total cost.

7. 6 napkins at $3.45 each _____   8. 8 napkins at $2.15 each _____

9. 4 placemats at $3.95 each _____   10. 6 towels at $1.29 each _____

Part Three. Compare costs. You may need another sheet of paper to do the work.

11. Fancy dinner napkins cost $7 each. Matching placemats cost $8 each. A package of 4 napkins and 4 mats costs $50.

   a. What is the total cost of the items if you buy them separately? _____

   b. How much can you save if you buy the package instead? _____

12. A tablecloth costs $18.95. Placemats are $4.75 each. Napkins are $3.50 each. A package contains: 1 tablecloth, 2 placemats, and 2 napkins. The package price is $29.98.

   a. What is the total cost of the items if you buy them separately? _____

   b. How much can you save if you buy the package instead? _____

Cost of Tools for Cooking and Serving

Fran has a new apartment. She wants to buy a set of kitchen knives priced at $25.00. But the total cost will be more than the amount on the price tag. She must also pay sales tax. Where Fran lives, there's a 5% sales tax.

Problems: a. How much is the sales tax?
                  b. What is the total cost?

Solutions:

a. $25.00 price
    ×   .05 percent of tax
       $1.25 sales tax

b. $25.00 price
   +   1.25 tax
     $26.25 total cost

Directions. Find the sales tax and final cost for each of the sets below. You may need another sheet of paper to work the problems.

---

1. **Chef's Utensils**

   $10.00
   5% tax   Sales tax $ _____   Total cost $ _____

---

2. **Glassware Set**

   $20.00
   5% tax   Sales tax $ _____   Total cost $ _____

---

3. **Casserole Set**

   $15.00
   3% tax   Sales tax $ _____   Total cost $ _____

---

4. **Dinnerware Set**

   $60.00
   3% tax   Sales tax $ _____   Total cost $ _____

---

5. **Cookware Set**

   $110.00
   4% tax   Sales tax $ _____   Total cost $ _____

Appliance Discounts

Sandy saw this ad for a blender she wanted to buy. It said "10% OFF."

Problems: a. How much will the discount save her?
              b. What price will Sandy pay?

Solutions:

a. $30.00   regular price
    ×   .10   percent of discount
    _____
    $3.00   amount saved

b. $30.00   regular price
    − 3.00   amount saved
    _____
    $27.00   final price

Directions. Find the savings and final prices for these kitchen appliances. You may need another sheet of paper to work the problems.

1. **MIXER**

   10% OFF
   Reg. $25   You save $ _____   You pay $ _____

2. **FOOD PROCESSOR**

   15% OFF
   Reg. $139   You save $ _____   You pay $ _____

3. **COFFEE MAKER**

   20% OFF
   Reg. $49   You save $ _____   You pay $ _____

4. **MICROWAVE OVEN**

   20% OFF
   Reg. $129   You save $ _____   You pay $ _____

5. **TOASTER OVEN**

   25% OFF
   Reg. $49   You save $ _____   You pay $ _____

Buying Applicances on Credit

Major kitchen appliances are expensive. Some stores let customers buy appliances on credit. This way, people can pay for their purchases over many months.

Example: Claudia paid $525 for this gas range. She paid for it in 25 monthly payments.

Problem: How much did she pay each month?

Solution: Divide the number of months into the total cost. cost.

$$25 \text{ months} \overline{)\begin{matrix} \$\ 21.00 & \text{monthly payment} \\ \$525.00 & \text{cost of range} \end{matrix}}$$

Claudia paid $21 per month for her new range.

Directions. Find out the monthly payments for each appliance below. You may need another sheet of paper to work the problems.

---

1. Dale bought as microwave oven for $180. He made 12 monthly payments. How much did he pay each month?

2. Betty's dishwasher cost $369. She paid for it in 18 months. How much did she pay each month?

---

3. An electric range cost $459. Cindy and Norman paid for it over a 25-month period. What was their monthly payment?

4. The Mullers bought a freezer for $270. If they had 24 months to pay, what did they pay per month?

---

5. Jan purchased a refrigerator for $949. She made 25 monthly payments. How much did Jan pay per month?

6. Lori's microwave cost $129. What was her monthly payment if she had 12 months to pay?

---

7. Kira bought a range for $649 and a dishwasher for $399. She had 25 months to pay. How much did she pay the store per month for both of her new appliances?

8. Mr. Young bought a new refrigerator for $1399 and a freezer for $299. If he paid for them in 24 monthly installments, what did he pay per month?

Review on Kitchen Supplies

Directions. Do the problems on another sheet of paper. Choose the *best* operation to solve each problem. Circle its letter in the correct column. Then write it in the blank in the riddle. One has been done for you.

Problems	+	−	×	÷	Answers
1. Lois bought a refrigerator for $550. She had 25 months to pay. How much did she pay each month?	E	G	L	(N)	# 22.00
2. A cookware set costs $75. Sales tax is $3. What is the final cost?	I	B	K	S	
3. Dish soap costs $2.20 and sponges cost $.98. How much for both?	N	A	O	J	
4. Cloth napkins cost $2.39 each. How much for 4 napkins?	T	H	Y	R	
5. A toaster costs $22. You can get 10% off. What amount can you save?	F	D	S	M	
6. A microwave costs $324. What is each monthly payment if you have 24 months to pay?	W	S	T	G	
7. A set of 6 placemats costs $17.88. What is the cost for one placemat?	N	Q	E	A	
8. Ron saved $12 when he bought a broiler oven. The regular price was $60. How much did Ron pay?	G	F	B	E	
9. Diane bought cleaning supplies for $1.49 and $3.29 and $2.99. How much did she spend in all?	P	X	A	U	
10. Stefan's mixing bowls cost $15.50 plus 4% sales tax. What amount was the tax?	H	F	L	C	

RIDDLE: What do jet pilots use for cookware?

__ __ __ __ _N_ __   __ __ __ __
8 10 4   2   1   6   9 7 3 5

© 1988, 1997 J. Weston Walch, Publisher   43   *Kitchen Math*

Computation Posttest: Part One

Directions. See how many problems you can do. You may need another sheet of paper to work the problems.

ADD

1.   315
     + 25
    _____

2.   265
        40
    + 175
    _____

3. $4.10
      $ .25
    + $2.60
    _____

4. $ 2.49
      $ .68
    + $11.40
    _____

5. $39.95
    + $ 1.99
    _____

6. $\frac{1}{4} + \frac{1}{4} =$ _____

7. $2\frac{1}{2}$
    + $\frac{1}{4}$
    _____

SUBTRACT

8.   2400
    − 1750
    _____

9. $13.00
    − $ 5.00
    _____

10. $2.79
    − $ .12
    _____

11. $10.00
    − $ 8.09
    _____

12. $55.00
    − $49.98
    _____

13. $.86
    − $.79
    _____

14. $3\frac{1}{4}$
    − $\frac{1}{4}$
    _____

15. $1\frac{1}{2}$
    − $\frac{1}{4}$
    _____

Kitchen Math

Computation Posttest: Part Two

Directions. See how many problems you can do. You may need another sheet of paper to work the problems.

MULTIPLY

16.
$$\begin{array}{r} 16 \\ \times\ 4 \\ \hline \end{array}$$

17.
$$\begin{array}{r} \$1.30 \\ \times\ \ \ 6 \\ \hline \end{array}$$

18.
$$\begin{array}{r} \$\ .89 \\ \times\ \ \ 5 \\ \hline \end{array}$$

19.
$$\begin{array}{r} \$1.48 \\ \times\ .25 \\ \hline \end{array}$$

20.
$$\begin{array}{r} \$\ 25 \\ \times\ .20 \\ \hline \end{array}$$

21. $\$199\ \times\ 10\%\ =$ _____

22. $\frac{1}{2}\ \times\ 8$ _____

DIVIDE

23. $8\,)\overline{64}$

24. $6\,)\overline{\$2.40}$

25. $60\,)\overline{180}$

26. $\frac{1}{4}\ \div\ 2\ =$ _____

Kitchen Math

Application Posttest: Part One

(For Masters #6–17)

Directions. See how many problems you can do. You may need another sheet of paper to work the problems.

1. A milk shake has 520 calories.
   A hamburger has 370 calories.
   French fries have 155.
   What are the total calories?

2. One serving of cheese gives you 110 calories. Suppose you should get 2300 calories in one day. How many calories will come from other foods?

3. Ross bought these foods:
   Steak $10.30
   Potatoes $ .60
   Mushrooms $ 2.00
   What was the total cost?

4. Pam's food budget is $35. She has spent $29. How much does Pam have left?

5. There are 18 guests. Each wants 6 crackers. How many crackers do they want in all?

6. A bag of 6 dinner rolls costs $1.29. Jonathan wants 4 bags. What will he pay?

7. Jim's groceries cost $8.75. He pays with a $10 bill. How much change will he get?

8. Bananas are $.35 per pound. How much will Chris pay for 4 pounds?

9. A package of fish weighs 2.5 pounds. The price is $3.50 per pound. What is the price of this package?

10. Carol sees 2/$.49 marked on canned juice. How much will she pay for just one can?

11. A small baby bottle holds 4 ounces. A can of ready-to-use formula contains 32 ounces. How many baby bottles will one can fill?

Kitchen Math

Application Posttest: Part Two

(For Masters #19–30)

Directions. See how many problems you can do. You may need another sheet of paper to work the problems.

---

12. Cereal costs $1.79. Sal has a coupon that says 25¢ off. What will he pay for juice?	13. Cheese is $1.75 at one store. It is $2.10 at another store. How much can you save at the less expensive store?

14. One brand of cream is $1.23 per pint. The other brand is $1.13. How much can you save buying 3 pints of the lower-priced brand?	15. Which is the better buy?
	  a. 32 oz for $1.92
	  b. 24 oz for $1.68

16. A gallon jug of vinegar costs $2.29. A quart costs $.99. How much can you save if you buy the jug instead of 4 quarts?	17. There are 16 ounces in one pint. How many ounces are in 3 pints?
18. You combine 550 mL of soda and 350 mL of cider. How much liquid do you have in all?	19. You combine 1¼ cups of white flour and ¾ cup of wheat flour. How much flour do you have in all?
20. Suzy has 3 cups of flour. If she uses ½ cup, how many cups are left?	21. A recipe uses 2 onions. How many onions make a half recipe?
22. Two lemons make 1 ounce of juice. How many lemons make 5 ounces?	

Name _____   Date _____

Application Posttest: Part Three

(For Masters #32–42)

Directions. See how many problems you can do. You may need another sheet of paper to work the problems.

23. Ted starts to make lunch at 11:15. It takes him 30 minutes. What time is he done?	24. Cook at high for 6½ minutes. Then cook at medium for 15 minutes. What is the total cooking time?
25. The meat needs to cook for 90 minutes. To be ready at 7:30, what time should it start cooking?	26. Your pan holds 6 biscuits. You need to cook 18 in all. How many batches of biscuits must you cook?
27. A roast cooks for 28 minutes per pound. Maria's roast is 5 pounds. How long will it take to cook?	28. Dale buys cleaning supplies on sale for $3.95. The regular cost would be $5.98. How much does Dale save?
29. A set of 6 dish towels costs $8.88. How much for one towel?	30. A set of dishes costs $52 plus 4% sales tax. What amount is the tax?
31. Cliff's toaster oven cost $65. He got a discount of 20% off the price. What amount did he save?	32. Ron buys a freezer for $270. He has 12 months to pay. How much will he pay each month?

Answer Key

Master #1 Computation Pretest: Part One

1. 450
2. 920
3. $3.95
4. $21.85

5. $52.04
6. 1
7. $3\frac{3}{4}$
8. 750

9. $5.00
10. $1.74
11. $2.94
12. $1.03

13. $.09
14. 2
15. $\frac{1}{4}$

Master #2 Computation Pretest: Part Two

16. 144
17. $10.50
18. $3.16
19. $.61

20. $7.00
21. $29.90
22. 3
23. 8

24. $.40
25. 2
26. $\frac{1}{4}$

Master #3 Application Pretest: Part One

1. 395
2. 2170
3. $11.90

4. $8
5. 64
6. $4.17

7. $.50
8. $2.34
9. $5.25

10. $.33
11. 4

Master #4 Application Pretest: Part Two

12. $.99
13. $.26
14. $.30
15. a. $.04 per oz;
    b. $.05 per oz
    The better buy is (a)

16. $.59
17. 40 ounces
18. 800 mL
19. $2\frac{1}{2}$ cups

20. $1\frac{1}{2}$
21. 2
22. 5 oranges

Master #5 Application Pretest: Part Three

23. 2:45
24. $22\frac{1}{2}$ min
25. 5:45

26. 3
27. 150 minutes
    (or 2 hrs 30 min or $2\frac{1}{2}$ hr)

28. $2.20
29. $2.25

30. $.60
31. $3.50
32. $17.50

Master #6   Counting Calories

1. 450
2. 220
3. 490
4. 1045
5. 580
6. 835
7. Meal #4 has the highest total calories.
8. Meal #2 has the lowest total calories.
9. Yes, Carla got enough calories.
10. Mark got too few calories (not nearly enough).

Master #7   Nutrition Facts

1. 40 calories
2. 95 percent
3. 90 percent
4. 26 grams
5. 273 grams

Master #8   Figuring the Cost of One Meal

1. Meat choices and costs will vary.
2. Other food choices and costs will vary.
3. Total costs will vary, but should not exceed $25.
   Students must add the total costs correctly.
4. Amounts remaining will vary, and should probably be less than a dollar. Students
   must subtract their totals from $25 correctly.

Master #9   The Food Budget

1. $13 left
2. $7 left
3. a. $38 spent   b. $12 left
4. a. $46 spent
   b. No, it was not more than the budget.
5. a. $46 spent
   b. Yes, it was more than the budget.
6. a. $370 spent
   b. $130 left

Master #10   Planning a Cookout

For 24 students:	How many to buy:
72 cookies	5 packages
48 hot dogs	4 packages
48 buns	6 bags
144 ounces (chips)	4 bags
192 ounces (drink)	6 cans
48 slices	4 melons

Master #11 Review on Planning Ahead

1.	L	(add)	200 calories	7.	K	(mult)	18 eggs
2.	U	(sub)	2100 calories	8.	L	(sub)	$2.75 more
3.	C	(mult)	36 rolls	9.	D	(div)	9 nuggets
4.	S	(div)	5 cookies	10.	P	(add)	$2.88 total
5.	E	(mult)	42 calories	11.	I	(sub)	$1.71 left
6.	L	(sub)	75 percent				

Riddle: DULL PICKLES

Master #12 The Shopping List

1. BETH'S LIST:

1.99	$1.99
.99	.99
2 × .99	1.98
1.59	+ 1.59
TOTAL:	$6.55

2. AMY'S LIST:

1.79	$1.79
1.99	1.99
2 × .99	1.98
.67	+ .67
TOTAL:	$6.43

3. JEFF'S LIST:

2 × .67	$1.34
2 × 3.59	7.18
2.29	2.29
2.99	+ 2.99
TOTAL:	$13.80

4. JUAN'S LIST:

.99	$ .99
1.49	1.49
3 × 2.29	6.87
2 × 1.59	+ 3.18
TOTAL:	$12.53

5. LEE'S LIST:

4 × 4.29	$17.16
3 × 1.99	5.97
2 × 1.99	3.98
2 × 2.99	+ 5.98
TOTAL:	$33.09

Master #13 Count Your Change

1. $1.00	5. $1.00	9. Yes
2. $.50	6. $6.54	10. The dog biscuits
3. $.30	7. $7.00	(costing $1.39) were taxable.
4. $.03	8. $.46	

Master #14 Fresh Produce by the Pound

1. $.90	5. $5.94	9. $3.15
2. $2.37	6. $5.38	10. $3.38
3. $1.95	7. $4.47	
4. $2.45	8. $2.25	

Master #15   Read the Label

1. $.79
2. $1.32
3. $.99
4. $3.27
5. $9.76

Master #16   How Much for One?

1. $.20
2. $.20
3. $.25
4. $.27
5. $.33
6. $.38
7. $.50
8. $.48
9. $.45
10. $.60

Master #17   Buying Baby Food

1. 5 days
2. 15 days
3. a. 8 bottles
   b. 4 days
4. a. 14 bottles
   b. $3\frac{1}{2}$ (or 3.5) days
5. a. $1.56
   b. $10.92
6. a. $1.85
   b. $12.95

Master #18   Review on Shopping for Food

1. A (add) $ 1.60
2. E (mult) $19.95
3. S (mult) $ 3.27
4. O (div) $ .25
5. O (mult) $3.96
6. P (sub) $2.50
7. T   (mult) $1.20
8. T   (div) $4.00
9. T   (sub) $7.11
10. A (div) $ .34
11. M (mult) $4.48

Riddle: TOMATO PASTE

Master #19   Saving Money with Coupons

1. $1.00
2. $2.10
3. $.84
4. $.67
5. $.90
6. 12 cents
7. No, the item's price is not given on the coupon.
8. No, the coupon may not be used after the expiration date.
9. $.83
10. $.88

Master #20   *Comparing Store Prices*

1. TUNA
   $1.14
   − .99
   $ .15 difference

2. MARGARINE
   $1.29
   − .55
   $ .74 difference

3. KETCHUP
   $1.40
   −1.29
   $ .11 difference

4. APPLES
   $1.39
   −1.29
   $ .10 difference

5. RAISINS
   $1.59
   − .38
   $ .21 difference

6. a. $8.06
   b. $7.04
   c. $6.39

7. Mammoth Foods has the lowest total price for the items, so Nancy could save some money buying her food there.
8. She could save $1.67 on these five items.

Master #21   *Comparing Brands*

1. WAX BEANS
   a. Major brand $ .79
   b. Generic   − .49
   c. Difference $ .30
   d. You could save $.60 on 2 cans.

2. PEACHES
   a. Major brand $1.15
   b. Generic   − .99
   c. Difference $ .16
   d. You could save $.64 on 4 cans.

3. PEARS
   a. Major brand $1.15
   b. Generic   −1.05
   c. Difference $ .10
   d. You could save $.30 on 3 cans.

Master #22   *Unit Pricing*

Note: Students may write their answers either as cents (5.5¢) or as dollars ($0.055).

1. a. 65¢ per carton
   b. 72.5¢ per carton
   The better buy is (a).

2. a. 28¢ for one
   b. 26.5¢ for one
   The better buy is (b).

3. a. 31¢ per lb
   b. 52.5¢ per lb
   The better buy is (a).

4. a. 56¢ per lb
   b. 54.5¢ per lb
   The better buy is (b).

5. a. 4.5¢ per oz
   b. 5.5¢ per oz
   The better buy is (a).

Master #23   Buying Large Quantities

1.   <u>2-lb bag–$.99</u>   <u>5-lb bag–$2.39</u>   <u>10-lb bag–$3.89</u>
      a. 10 bags   4 bags   2 bags
      b. cost $9.90   cost $9.56   cost $7.78
      c. save $3.41   save $3.07   save $1.29

2. a. cost $4.18
      b. save $.39

3. a. cost $7.45
      b. save $1.46

Master #24   Review on Stretching Food Dollars

1. F   (sub) $1.85
2. L   (sub) $.04
3. S   (div) $.18
4. T   (div)   4 bags
5. W (mult) $1.35
6. I   (mult) $2.36

7. O   (sub) $.56
8. F   (div) $ .25 (4/$1) is a better
                   buy than $.28 (3/$.84).
9. E   (sub) $1.59
10. S (sub)   16.5¢ (or $.165)
11. B (div)   3¢ (or $.03)

Riddle: STIFF ELBOWS

Master #25   Equal Measures

1. 4 squares
2. 32 ounces (or 1 quart)
3. 9 teaspoons
4. 8 tablespoons (or 1 stick)
5. 12 cups

6. 16 ounces (or 1 pound)
7. 6 teaspoons
8. 8 cups
9. 8 quarts
10. 8 ounces (16 T = 1 cup = 8 oz)

Master #26   Combining Liquid Ingredients

1. 750 mL
2. 300 mL
3. 1000 mL (or 1 liter)
4. 950 mL
5. 1500 mL (or 1.5 liters)

6. 4 liters
7. 1350 mL (or 1.35 liters)
8. 625 mL
9. 1500 mL (or 1.5 liters)
10. 5.5 liters

Master #27   Combining Dry Ingredients

1. $3\frac{1}{2}$ cups
2. $3\frac{1}{4}$ cups
3. $2\frac{2}{3}$ cups
4. $4\frac{1}{2}$ cups
5. $\frac{3}{3} = 1$ cup

6. $2\frac{3}{4}$ cups
7. $\frac{2}{2} = 1$ cup
8. $4\frac{1}{2}$ cups
9. $2\frac{2}{4} = 2\frac{1}{2}$ cups
10. $1\frac{6}{4} = 2\frac{2}{4} = 2\frac{1}{2}$ cups

Master #28　*Figuring Leftovers*

1. $\frac{1}{2}$ lb
2. $1\frac{3}{4}$ lb
3. $4\frac{1}{2}$ lb
4. $\frac{1}{2}$ cup

5. 1 cup
6. $2\frac{1}{4}$ cups
7. $2\frac{1}{2}$ cups

8. $\frac{3}{4}$ lb
9. 3 lb
10. $\frac{1}{2}$ cup

Master #29　*Adjusting a Recipe*

Italian Meatballs	Half Recipe 2 servings	Double Recipe 8 servings
Dry bread:	$1\frac{1}{2}$ slices	6 slices
Ground beef:	$\frac{3}{4}$ pound	3 pounds
Eggs:	1 egg	4 eggs
Cheese:	$\frac{1}{4}$ cup	1 cup
Butter:	$\frac{1}{2}$ tablespoon (or $1\frac{1}{2}$ t)	2 tablespoons
Onion:	2 tablespoons	8 tablespoons (or $\frac{1}{2}$ cup)
Parsley:	1 tablespoon	4 tablespoons
Garlic:	$\frac{1}{2}$ clove	2 cloves
Oregano:	$\frac{1}{4}$ teaspoon	1 teaspoon
Salt:	$\frac{1}{8}$ teaspoon (a pinch)	$\frac{1}{2}$ teaspoon

Master #30　*Ratios in the Kitchen*

1. $\frac{1}{2}=\frac{?}{8}$ (4)
2. $\frac{3}{4}=\frac{?}{8}$ (6)
3. $\frac{2}{1}=\frac{?}{3}$ (6)

4. $\frac{1}{5}=\frac{2}{?}$ (10)
5. $\frac{3}{1}=\frac{9}{?}$ (3)
6. $\frac{2}{2}=\frac{6}{?}$ (6)

7. 8 oranges
8. 12 ounces
9. 6 slices
10. 10 cups

Master #31 *Review on Using Measurements*

1. O　(add)　900 mL
2. A　(mult)　48 ounces
3. N　(sub)　$1\frac{1}{2}$ pounds
4. S　(div)　$1\frac{1}{2}$ cups
5. D　(add)　$\frac{2}{4}=\frac{1}{2}$ cup
6. U　(add)　4.5 liters

7. A　(add)　$1\frac{3}{4}$ cups
8. O　(mult)　160 ounces
9. R　(mult)　6 pounds
10. F　(sub)　$2\frac{1}{4}$ cups
11. L　(mult)　3 cups

Riddle: LOAFS AROUND

Master #32   Cooking Times

1. 2:20	5. 4:30	8. 6:35
2. 12:05	6. 10:10	9. 6:45
3. 6:50	7. 2:30	10. 9:05
4. 10:15		

Master #33   Microwave Cooking Times

1. Press: 3 0
2. Press: 3 0 0
3. Press: 1 4 5
4. Press: 1 2 3 0
5. 22 min
6. $23\frac{1}{2}$ min (or 23 min 30 sec)
7. $22\frac{1}{2}$ min (or 22 min 30 sec)
8. 8 min
9. 38 min
10. 48 min

Master #34   Serving a Meal on Time

1. 6:00	5. 11:35	8. 5:30
2. 5:00	6. 12:45	9. 12:05
3. 4:15	7. 5:00	10. 6:10
4. 5:30		

Master #35   Cooking in Batches

1. 4 batches; 12 minutes
2. 3 batches; 60 minutes
3. 2 batches; 24 minutes
4. 5 batches; 20 minutes
5. 2 batches; 30 minutes

(*Note:* The final problem is tricky but realistic. Since only 24 of the 36 biscuits can be baked at once, the remaining 12 biscuits will be done as a separate batch which must bake for the full 15 minutes.)

Master #36   How Long to Cook Meat?

1. 6 lb × 20 min = 120 min = 2 hours
2. 5 lb × 20 min = 100 min = 1 hour and 40 min
3. 7 lb × 25 min = 175 min = 2 hours and 55 min
4. 8 lb × 28 min = 224 min = 3 hours and 44 min
5. 4 lb × 22 min = 88 min = 1 hour and 28 min

(*Note:* Cooking times shown on the chart have been simplified for the exercise and are approximate times only. For more accurate cooking times, consult your cookbook.)

Master #37   Review on Preparation Time

1. E   (div) $2\frac{1}{2}$ hours
2. O   (add)   3:15
3. E   (div)   3 batches
4. N   (mult)   75 minutes (or 1 hour and 15 min)
5. A   (mult)   45 minutes
6. P   (sub)   12:10
7. C   (mult)   88 minutes (or 1 hour and 28 min)
8. G   (add)   6:00
9. S   (add) $13\frac{1}{2}$ minutes
10. K   (sub)   6:35

Riddle: SPONGE CAKE

Master #38 Buying Cleanup Supplies on Sale

1. Ann:

	Reg. prices	Sale prices
	$1.49	$1.37
	$2.19	$1.89
	+$2.37	+$1.99
	$6.05	$5.25

Ann can save $.80.

2. Tom:

	Reg. prices	Sale prices
	$3.29	$2.99
	$1.25	$ .99
	+$.89	+$.78
	$5.43	$4.76

Tom can save $.67.

3. Marya:

	Reg. prices	Sale prices
	$1.79	$1.49
	$3.19	$2.98
	+$1.49	+$1.37
	$6.47	$5.84

Marya can save $.63.

4. Liz:

	Reg prices	Sale prices
	$2.19	$1.89
	$.73	$ .69
	+$1.25	+$.99
	$4.17	$3.57

Liz can save $.60.

Master #39   Buying Kitchen and Table Linens

1. $1.74
2. $1.48
3. $3.75
4. $4.95

5. $1.49
6. $1.79
7. $20.70

8. $17.20
9. $15.80
10. $7.74

11. a.
$$\begin{array}{r} \$7 \times 4 = \$28 \\ + \$8 \times 4 = \$32 \\ \hline \text{total:} \quad \$50 \end{array}$$

   b.
$$\begin{array}{r} \$60 \\ - \$50 \\ \hline \text{save:} \quad \$10 \end{array}$$

12. a.
$$\begin{array}{r} \$18.95 \\ \$4.75 \times 2 = \$ \ 9.50 \\ + \$3.50 \times 2 = \$ \ 7.00 \\ \hline \text{total:} \quad \$35.45 \end{array}$$

   b.
$$\begin{array}{r} \$35.45 \\ - \$29.98 \\ \hline \text{save:} \quad \$5.47 \end{array}$$

Master #40   Cost of Tools for Cooking and Serving

1. sales tax $.50; total cost $10.50
2. sales tax $1.00; total cost $21.00
3. sales tax $.45; total cost $15.45
4. sales tax $1.80; total cost $61.80
5. sales tax $4.40; total cost $114.40

Master #41   Appliance Discounts

1. save $2.50; pay $22.50
2. save $20.85; pay $118.15
3. save $9.80; pay $39.20
4. save $25.80; pay $103.20
5. save $12.25; pay $36.75

Master #42   Buying Appliances on Credit

1. $15.00
2. $20.50
3. $18.36
4. $11.25
5. $37.96
6. $10.75
7. Monthly payment for both was $41.92
8. Monthly payment for both was $70.75

Master #43   Review on Kitchen Supplies

1. N (div) $22.00
2. I (add) $78.00
3. N (add) $3.18
4. Y (mult) $9.56
5. S (mult) $2.20
6. G (div) $12.50
7. A (div) $2.98
8. F (sub) $48.00
9. P (add) $7.77
10. L (mult) $.62 (or 62¢)

Riddle: FLYING PANS

Master #44   *Computation Posttest:* *Part One*

1. 340
2. 480
3. $6.95
4. $14.57
5. $41.94
6. $\frac{1}{2}$
7. $2\frac{3}{4}$
8. 650
9. $8.00
10. $2.67
11. $1.91
12. $5.02
13. $.07
14. 3
15. $1\frac{1}{4}$

Master #45   *Computation Posttest:* *Part Two*

16. 64
17. $7.80
18. $4.45
19. $.37
20. $5.00
21. $19.90
22. 4
23. 8
24. $.40
25. 3
26. $\frac{1}{8}$

Master #46   *Application Posttest:* *Part One*

1. 1045
2. 2190
3. $12.90
4. $6
5. 108
6. $5.16
7. $1.25
8. $1.40
9. $8.75
10. $.25
11. 8

Master #47   *Application Posttest:* *Part Two*

12. $1.54
13. $.35
14. $.30
15. a. $.06 per oz;
   b. $.07 per oz
   The better buy is (a)
16. $1.67
17. 48
18. 900 mL
19. 2 cups
20. $2\frac{1}{2}$
21. 1
22. 10 lemons

Master #48   *Application Posttest:* *Part Three*

23. 11:45
24. $21\frac{1}{2}$ min
25. 6:00
26. 3
27. 140 minutes (or 2 hrs 20 min)
28. $2.03
29. $1.48
30. $2.08
31. $13.00
32. $22.50

NOTES

NOTES

NOTES

NOTES